ARMORED DINOSAURS

BY "DINO" DON LESSEM
ILLUSTRATIONS BY JOHN BINDON

LERNER PUBLICATIONS COMPANY / MINNEAPOLIS

To Ken Carpenter, armored dinosaur expert

Reprinted in 2006

This book is available in two editions:
Library binding by Lerner Publications Company,
a division of Lerner Publishing Group
Soft cover by First Avenue Editions,
an imprint of Lerner Publishing Group
241 First Avenue North
Minneapolis, MN 55401 U.S.A.

Website address: www.lernerbooks.com

Library of Congress Cataloging-in-Publication-Data

Lessem, Don.
Armored dinosaurs / by Don Lessem ; illustrations by John Bindon.
p. cm. — (Meet the dinosaurs)
Includes index.
Summary: Describes how armor helped dinosaurs such as *Albertosaurus* and *Edmontonia* to survive attack, as well as how scientists have learned about armored dinosaurs through studying fossils.
ISBN-13: 978-0-8225-1374-2 (lib. bdg. : alk. paper)
ISBN-10: 0-8225-1374-9 (lib. bdg. : alk. paper)
ISBN-13: 978-0-8225-2570-7 (pbk. : alk. paper)
ISBN-10: 0-8225-2570-4 (pbk. : alk. paper)
1. Ornithischia—Juvenile literature. (1. Ornithischians. 2. Dinosaurs.) I. Bindon, John. II. Title.
QE862.O65L48 2005
567.915—dc21 2002154234

Printed in China
2 3 4 5 6 7 - DP - 11 10 09 08 07 06

TABLE OF CONTENTS

MEET THE ARMORED DINOSAURS

WELCOME, DINOSAUR FANS!

I'm "Dino" Don. Dinosaurs are my favorite animals. The armored dinosaurs had strong armor for protection. Here are some fast facts on the armored dinosaurs you'll meet in this book. Have fun!

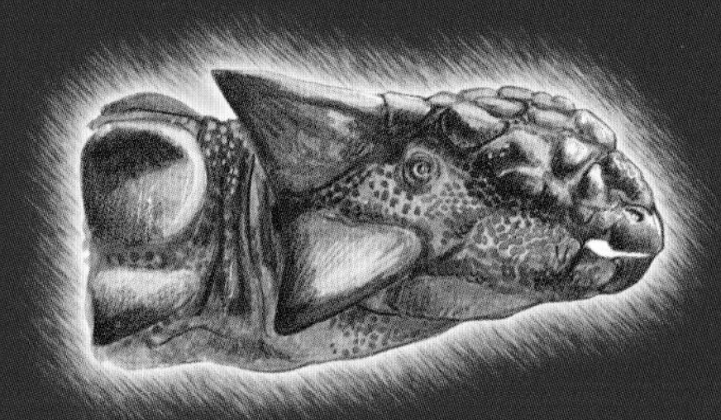

ANKYLOSAURUS **(AN-ky-luh-SAWR-uhs)**
Length: 25 feet
Home: western North America
Time: 65 million years ago

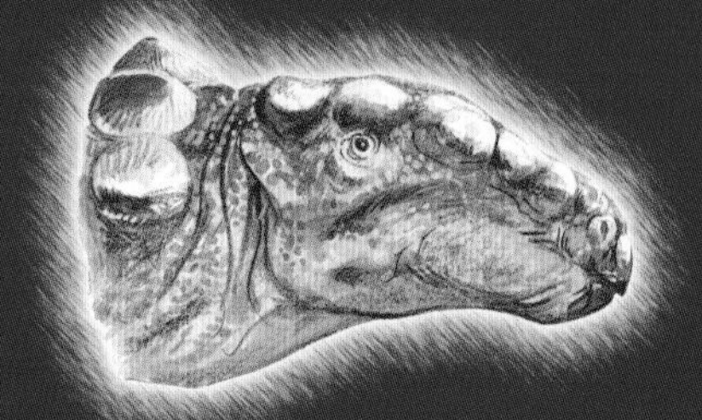

EDMONTONIA **(EHD-muhn-TOH-nee-uh)**
Length: 23 feet
Home: western North America
Time: 75 million years ago

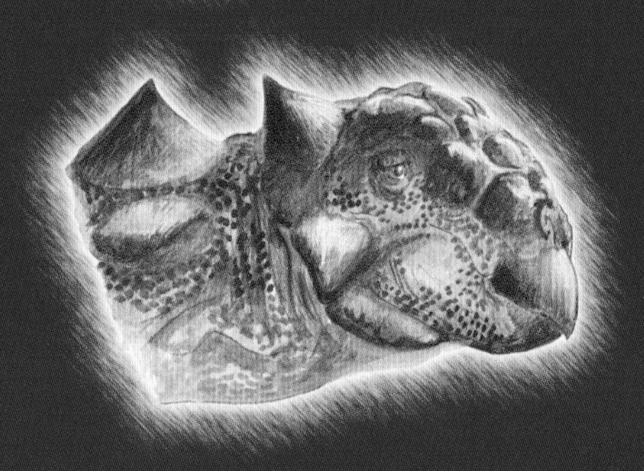

EUOPLOCEPHALUS **(YOO-oh-pluh-SEHF-uh-LUHS)**
Length: 17 feet
Home: western North America
Time: 75 million years ago

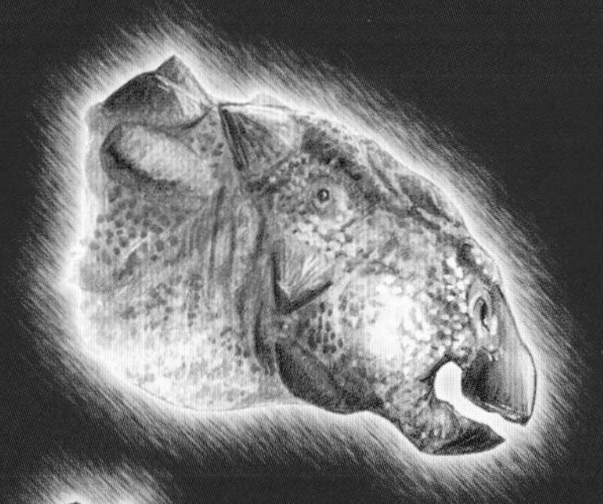

GASTONIA **(gas-TOH-nee-uh)**
Length: 20 feet
Home: western North America
Time: 130 million years ago

HUAYANGOSAURUS **(HWAH-yahng-uh-SAWR-uhs)**
Length: 13 feet
Home: eastern Asia
Time: 170 million years ago

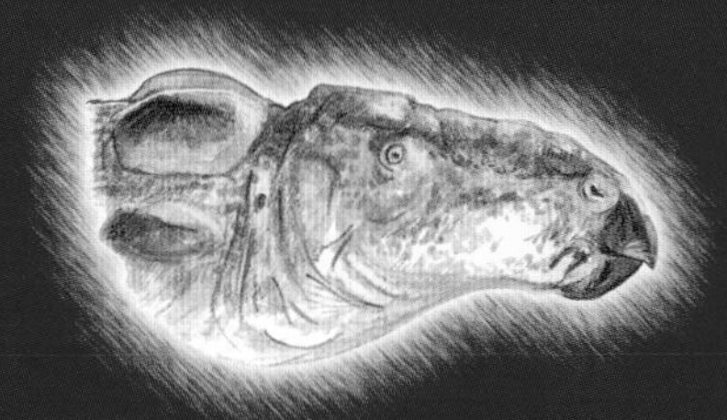

MINMI **(MIHN-mee)**
Length: 10 feet
Home: Australia
Time: 110 million years ago

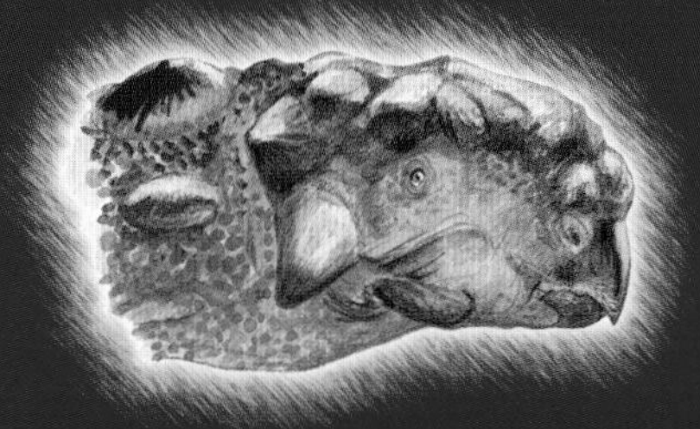

PINACOSAURUS **(PIHN-ak-uh-SAWR-uhs)**
Length: 18 feet
Home: eastern Asia
Time: 80 million years ago

STEGOSAURUS **(STEHG-uh-SAWR-uhs)**
Length: 25 feet
Home: western North America
Time: 145 million years ago

SAVED BY ARMOR

Here comes a hungry *Albertosaurus!* It's much bigger and faster than *Edmontonia.* But *Edmontonia* is covered with bony plates and spikes. *Albertosaurus* cannot bite or scratch through this thick **armor**.

THE TIME OF THE ARMORED DINOSAURS

Edmontonia, *Albertosaurus*, and other dinosaurs lived on land millions of years ago. Dinosaurs were like reptiles in some ways. They laid eggs, as snakes and other reptiles do. But dinosaurs weren't reptiles. Birds are closer relatives of dinosaurs than reptiles are.

One group of dinosaurs had armored bodies. Scientists think that armor helped protect these dinosaurs from **predators**. Predators are animals that kill and eat other animals. The armored dinosaurs weren't predators. They ate only plants.

DINOSAUR FOSSIL FINDS

The numbers on the map on page 11 show some of the places where people have found fossils of the dinosaurs in this book. You can match each number on the map to the name and picture of the dinosaurs on this page.

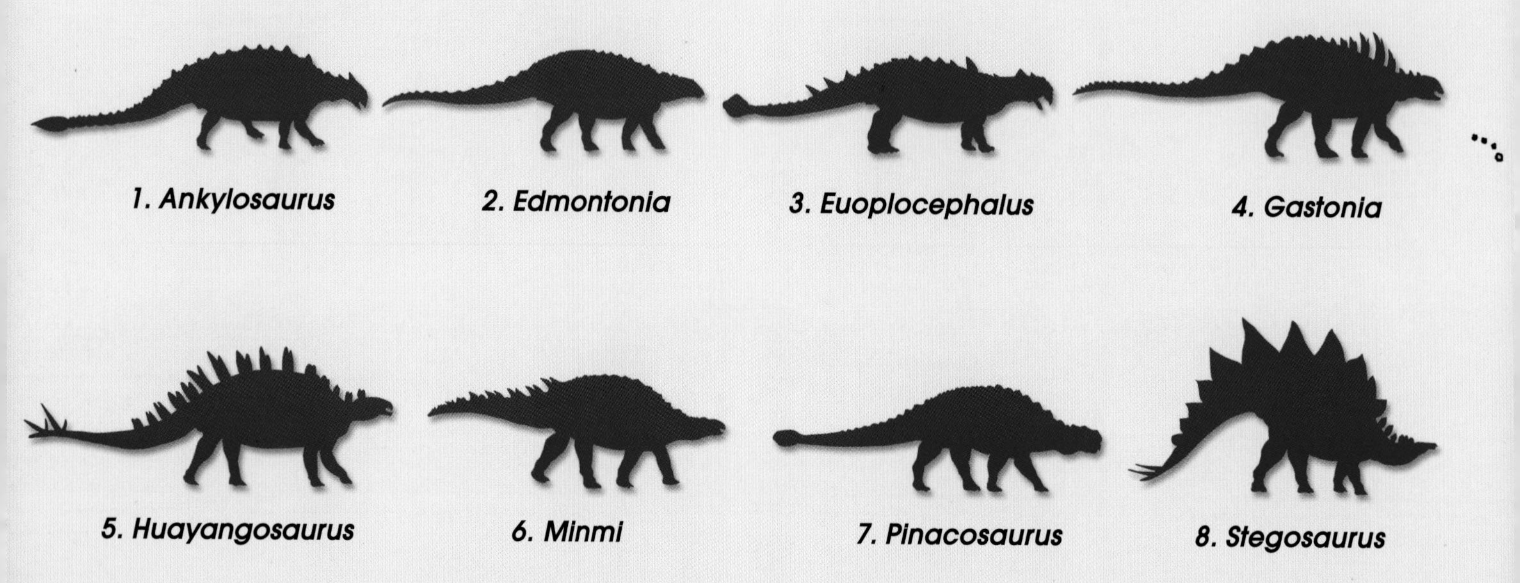

We know about armored dinosaurs from the traces they left behind, called **fossils**. Bony plates, spikes, and footprints help scientists understand how armored dinosaurs were built. But fossils can't tell us what color a dinosaur was. They can't tell us how its skin and muscles looked.

Still, fossils are our best clues to the puzzle of how armored dinosaurs might have lived. People have found fossils of these strange animals all over the world.

PLATES, SPIKES AND CLUBS

The deadly predator *Utahraptor* is attacking. It slashes at *Gastonia* with sharp claws that are as long as a carving knife. But *Gastonia* has the heaviest armor of any dinosaur. *Utahraptor* cannot slice through it. In time, the tired predator gives up.

Meat-eating dinosaurs were faster and smarter than armored dinosaurs. But spikes and thick plates kept armored dinosaurs safe from attack. Even their eyelids were covered with bone!

The giant back plates of *Stegosaurus* may have scared off some predators. But the plates were probably built more for show than for protection. This big male shows off his plates to attract a female.

Back plates may have also helped heat and cool *Stegosaurus*. The plates may have soaked up the sun's heat when *Stegosaurus* was cold. And they may have given off heat when it was hot. Heating and cooling probably wasn't easy for an animal the size of a garbage truck!

Euoplocephalus swishes its big, armored tail club. Any animal would be careful around such a weapon. But these *Pachycephalosaurus* are peaceful plant eaters. Why is the club aimed at them?

The dome-headed visitors want to munch on the plants that *Euoplocephalus* is eating. Armored dinosaurs could reach only plants that grew near the ground, like herbs and ferns. A low, fast-moving tail club may have helped some armored dinosaurs protect their meals.

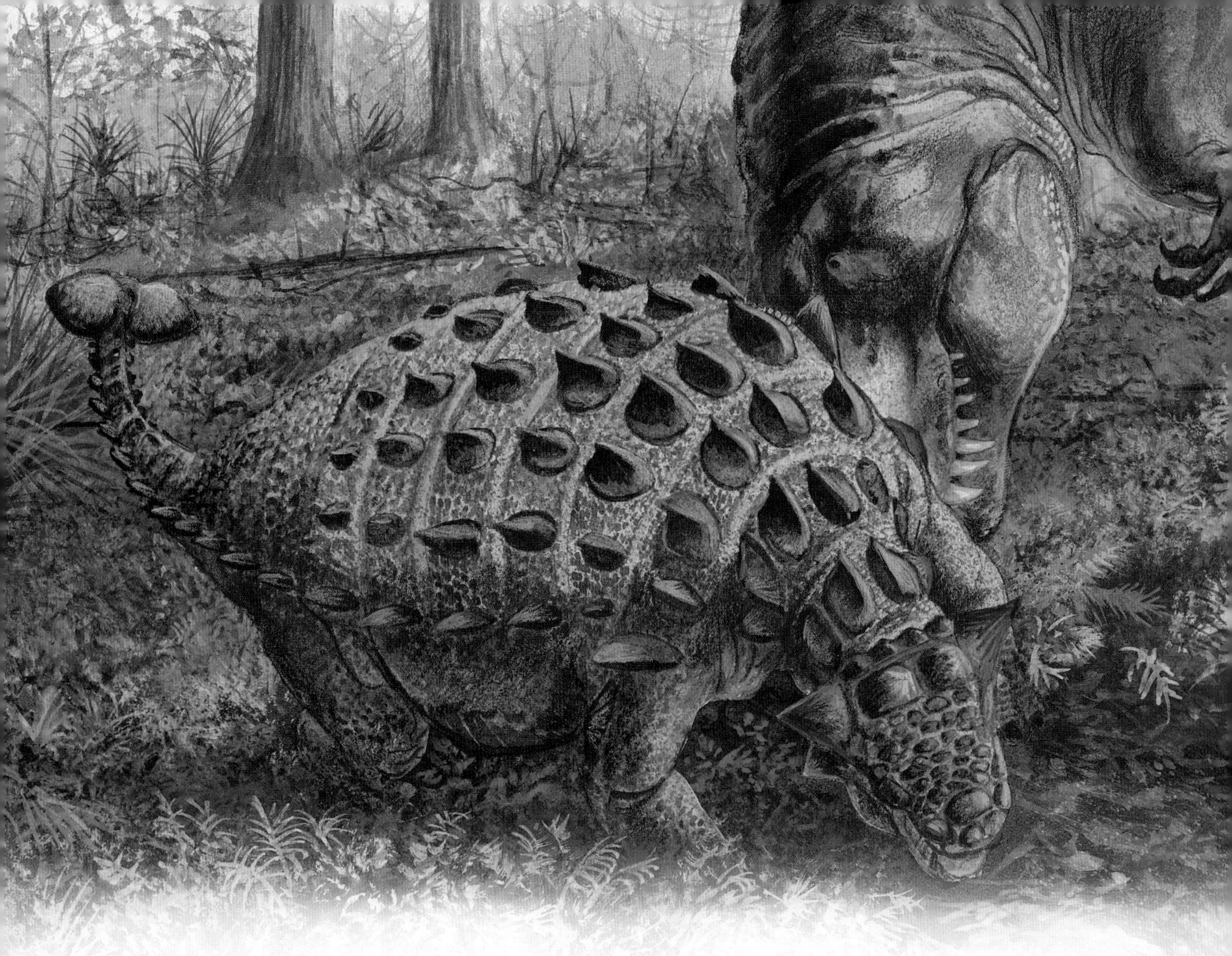

Can back spikes and body armor save *Ankylosaurus* from *Tyrannosaurus rex?* Maybe. But *Ankylosaurus* has no armor on its belly. The *T. rex* tries to flip *Ankylosaurus* to reach this soft flesh.

Ankylosaurus squats down low on its powerful legs. This armored giant weighs as much as four cars. It is so heavy and so close to the ground that *T. rex* cannot flip it over. Squatting probably helped many armored dinosaurs survive attack.

LIFE AND DEATH

How were armored dinosaurs born? Scientists think they hatched from eggs, the way other dinosaurs did. Armored dinosaurs were very small when they hatched.

We don't know if newborns like this *Minmi* had armor. If not, predators probably found them easy to kill. Armored mothers may have nested in groups. That way, they could protect the eggs and newborns.

A young *Huayangosaurus* has just hatched. It wanders from its nest. Nearby, two hungry *Gasosaurus* wait to snap it up. The newborn's parents and other adults arrive just in time. Together they scare off the *Gasosaurus*.

Huayangosaurus had no tail club to defend itself. Its back plates were not very big. But a group of these dinosaurs might still have been able to scare away large predators.

Armor could not always keep dinosaurs safe. These young *Pinacosaurus* huddle together beneath a sand dune. They are trapped in a blowing sandstorm.

Armored dinosaurs faced many dangers besides hungry meat eaters. Floods and storms ended the lives of many. But armored dinosaurs lived successfully all over the world for 100 million years.

ARMORED DINOSAUR MYSTERIES

The last armored dinosaurs died out suddenly 65 million years ago. So did the rest of the dinosaurs. What happened? Scientists are still trying to solve this mystery. Many think that an object from space called an **asteroid** smashed into the earth.

The asteroid's crash with the earth may have started fires and made volcanoes explode. Dust may have filled the air and blocked out sunlight. These changes would have killed dinosaurs, other animals, and many plants.

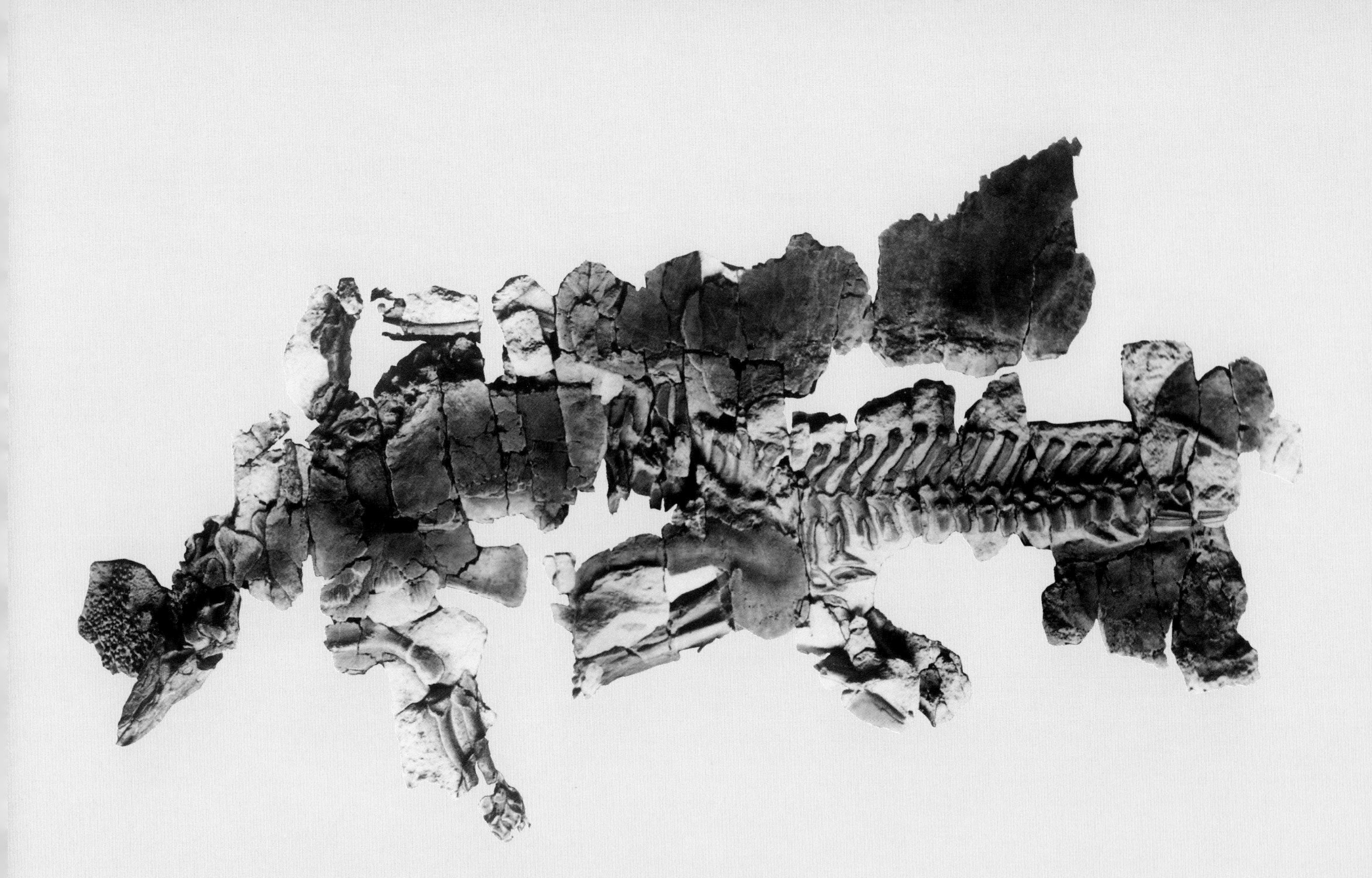

The story of *Stegosaurus* is another famous armored dinosaur mystery. People first found *Stegosaurus* fossils more than 100 years ago. Some bones were missing from the skeleton. The back plates lay flat on the ground. They were not joined to the skeleton's back.

Scientists could not tell how the plates had fit on the body of a living *Stegosaurus.* Did it have one row of plates or two? For years, scientists built models of *Stegosaurus* in different ways. Artists drew it in different ways too.

Scientists in Colorado found new clues to the *Stegosaurus* mystery during the 1990s. They dug up skeletons of *Stegosaurus* with almost every bone still in place. These fossils showed that *Stegosaurus* had two rows of plates on its back.

The *Stegosaurus* mystery has been solved. But people will always want to learn more about armored dinosaurs such as *Stegosaurus* and this *Gastonia*. They are among the strangest, most interesting animals that have lived on the earth.

GLOSSARY

armor (AR-mur): bony plates and spikes on the bodies of some dinosaurs

asteroid (AS-tur-oyd): a large rocky lump that moves in space

fossils (FAH-suhlz): the remains, tracks, or traces of something that lived long ago

predators (PREH-duh-turz): animals that hunt and eat other animals

INDEX